AGILE PROJECT MANAGEMENT

GUIDE TO LEADING AGILE PROJECT MANAGEMENT

Table of Contents

Introduction

The Agile Manifesto outlines the core values that an organization should embrace. These values emphasize collaboration and authentic human interactions. In addition, the methodology is designed to enhance technical excellence and foster sustainable development.

An agile team is comprised of developers, business people and designers who work together in a collaborative environment. They are empowered to be experts in their area and make rapid, yet informed decisions. To succeed with an Agile project, a team must understand how to manage its resources. Those resources include the time and budget. There are several methods and tools used in Agile project management, and students will learn the best practices.

During an Agile project, a team works on stories within the team's WIP limit. Small, frequent increments allow the team to gather feedback, and allow the team to accept a subset of the product's features.

Key Principles

Agile methodologies emphasize communication, collaboration and self-organizing teams. These techniques also prioritize the development of regular releases. They encourage teams to make decisions quickly, and focus on improving processes as they progress.

One key advantage of these practices is that they allow teams to focus on the value of the end product instead of the duration of the project. For example, if a team is not ready to ship working software, there is no way to measure its progress.

In addition, agile methods focus on streamlining data and automation, which reduces time spent on manual work. Another benefit is that they allow for frequent delivery of other products.

To ensure the successful delivery of a product, a business and engineering team must work together through every stage of the development process. The customer must also be involved throughout.

It is important to create a culture of trust and openness, and a commitment to teamwork. Ultimately, employees who are valued and empowered are more likely to respond to challenges and meet goals.

While a rapid release schedule can be taxing, it's often better to ship useful software now than to try to make a perfect product later.

The Agile Methodologies

Agile is a term used for a number of methodologies, which are designed to make a development team more efficient and effective. Its initial focus was on reducing the time to market while delivering high-quality code. To do this, it encouraged engineers to explore their craft. This meant that they should respond quickly to customer requirements.

In addition, the software development community began changing its approach to planning. This included the introduction of daily meetings and quick decisions. By allowing for continuous change, it helped to keep a team's focus.

As Agile has become more widely used outside of the Tech world, some people have expressed concern that the potency of the method will wane over time. Others have noted that cultural changes are often the biggest obstacle to adoption.

The Agile Project Management benefits and limitations

Agile project management is a great tool for delivering value and minimizing risks. Some businesses might find it difficult to adapt to a new style of management. Agile methods are characterized by small incremental releases and an emphasis on flexibility. These features help teams avoid risk and deliver products to the market quickly.

While the agile method provides a number of benefits, it is not right for every software project. Typically, the method is more effective for small-to-medium-sized companies and product-oriented organizations.

Teams have to work together and absorb feedback. This means that they must think creatively. During the implementation phase, an agile department can fix problems rapidly. They can also use client feedback to improve the next iteration. One of the more important factors of an agile approach is its ability to reduce go-to-market time. This means that the product owner can take advantage of a market opportunity before the competition.

Agile Project Management Frameworks

There is many different Agile Project Management Frameworks to choose from. Some of these include Test-Driven Development, Lean, Bimodal, XP, and Extreme Programming. Using the most appropriate one is important.

When you are choosing a framework, you need to consider what will be best for your team and your specific project needs. It breaks projects into smaller milestones, called sprints. These sprints are iterative and allow teams to make course corrections and learn quickly. The important part of an agile methodology is the ability to respond to change. Changes in your budget or timeline can happen at any time, and you need to be able to accommodate them.

Regardless of the specific approach you choose, it is a good idea to spend a few days communicating face-to-face.

Scrum - The Three Pillars of Scrum

Scrum is a team-based, iterative and incremental development method. The methodology is based on three pillars: product, process and people. In Scrum, a sprint is typically 30 days. However, a longer sprint can be accomplished in shorter spans.

The task board is an important part of the Scrum method. It allows teams to measure progress and improve features in response to user feedback. In the Scrum methodology, the Product Owner is responsible for prioritizing work based on the value it brings to the product or service.

Sprint planning is another essential step in the Scrum process. In this meeting, the Product Owner explains the features for the next iteration and how the team will achieve it.

Release planning is also a part of the Scrum process. Usually, release planning is done quarterly. During this event, the team validates the system's readiness for release and makes adjustments if necessary.

An interesting feature of the Scrum method is that it uses collaborative decision-making. This allows the team to quickly scale modules, improve products in response to user feedback and focus on the core functionalities.

How to Implement Kanban in Your Business

It promotes continuous improvement, improved flow, and greater predictability. Whether you use it in your engineering, manufacturing, or software development projects, it can help you deliver more quickly and with less risk.

Kanban uses limits on the number of work items to ensure that your teams are completing the most important tasks in a timely fashion. This helps reduce cycle time and lead times. Using a kanban board, you can identify bottlenecks in your production process and address them before they become a problem.

The Kanban method also promotes communication and information sharing. It allows teams to visualize their processes. Teams can then collaboratively determine changes that need to be made.

To implement a kanban system, you need to define what constitutes a finished task and what is in progress. When a new person joins the team, he or she may be given a working limitation on how much of the process he or she can complete.

Kanban helps improve the flow of work, eliminate waste, and increase quality. Compared to traditional methods, kanban is more flexible and does not require changes to the organization.

What is Lean?

It starts from the people doing the work, but it also involves improving the environment where the work is performed. As a result, companies become more innovative, competitive, and sustainable. These principles include mapping the value stream, defining value, and pursuing perfection. Value is defined as the value created for the customer. This value can be assessed from the perspective of the customer and it can also determine a buyer's willingness to pay for a product.

By analyzing the entire process, a company can develop a value-stream map. The value stream map outlines the life cycle of the product, capturing the flow of materials and information.

When a value stream map is completed, the company can create a system that focuses on efficient manufacturing. It can also be used to identify and correct any loopholes that can lead to waste.

Another important element of lean is the focus on employees. Lean managers should act as coaches for other workers and help them identify and fix problems. They can encourage them to make daily improvements and help them learn through experimentation.

During the process, the company should reward employees for their progress and milestones. In return, they will want to keep working with the company to continue implementing changes.

Types of Hybrid Frameworks

Hybrid frameworks are a great way to get apps designed quickly and easily. Users want to have instant access to information. They also want apps that load quickly. It is important to choose the right app framework to make sure your mobile applications perform well on all devices.

There are many types of hybrid frameworks to choose from. Some of them are NativeScript, jQuery, Flutter, and React Native. Each framework has its own unique benefits. You should consider your needs and your budget before selecting a framework.

In order to build a multi-platform app, you can use Xamarin. This mobile development framework is owned by Microsoft and provides native functionality. Developers can write the same code and deploy it on different operating systems, resulting in a better user experience. However, Xamarin is not the best choice for complex applications.

Another option for building cross-platform apps is Ionic. Unlike other frameworks, this framework allows you to write apps in multiple languages. Moreover, this framework is well supported by Adobe Air, Cordova, and PhoneGap.

Right Framework

Choosing the right framework for your project is a critical step towards completing a successful endeavor. It helps you streamline your development efforts, ensuring that your project churns out results and not disasters. A poorly chosen framework could mean lost time and resources, so make sure to do your homework.

The best frameworks should offer you a range of features and functionality. If you want to expand your app's capabilities, you'll need to look for a framework that will cater to a variety of devices. You'll also need to check the documentation for your chosen framework to make sure that it's able to do what you need it to.

When deciding on the right framework for your project, you'll need to consider a few important aspects. One of these is whether your application will be distributed across multiple servers. This means that you'll need to pick a framework that offers compatibility with different operating systems, as well as a range of programming languages. Another consideration is the amount of memory needed to run your application.

The Tools and Techniques

Managing an agile project can be a bit daunting at first, but it does offer some benefits. For starters, an Agile environment is very flexible. It can adapt to a change in scope. Another advantage is a focus on user engagement. Using an agile methodology, you can easily get feedback on a new feature or functionality.

Traditional approaches are often too rigid to accommodate changes in scope. They also depend on predictable tools and experience. Using two different methods on the same project can be counterproductive.

The Agile method has a few advantages, including improved teamwork. One benefit is that it allows the use of improvisations at every stage. By using an iterative approach, you can reduce costs and waste. The PERT method is another common project management method. Other notable Agile management techniques include incorporating lean and waterfall methodologies.

Guidelines for Writing User Stories and Acceptance Criteria

User stories and acceptance criteria are important elements in the development process. These documents are essential to ensure the client and development team have a common understanding of what to expect. The best user story and acceptance criteria are the ones that are clear, concise, and easy to understand. These documents can help reduce rework and deliver a product with a higher quality level. However, writing these can be difficult.

First, don't try to write acceptance criteria all in one go. You should consider writing these in small pieces over a period of time. Make sure your user stories and acceptance criteria are focused on what they need, not what they want.

Second, make sure your criteria are measurable. For example, you could include the number of days it will take to deliver the feature or the amount of code to be written. A metric can help you estimate how long your development will take and will give you an idea of whether or not you're on track.

Finally, avoid overly specific or overly broad criteria. Overly specific criteria can hinder your ability to maneuver. In addition, a very specific acceptance criterion might miss out on other user behaviors that are not accounted for.

Planning Poker and Estimation

Planning poker and estimation is a common tool used by agile development teams to determine the time needed to complete their tasks. It is a useful tool for prioritizing agile roadmap items. There are several techniques used in this estimation method, but the most popular is the Fibonacci sequence.

The process of planning poker and estimation involves the use of physical cards that are used to estimate the story points of the user story. A team member may make an estimate of the story points that will be required, but other members may disagree with this. This encourages communication, collaboration, and team building.

Although planning poker and estimation can be a very effective tool for estimating the time it will take for an agile team to complete its tasks, it is important to remember that there are limits to the effectiveness of this method. For example, it is best not to estimate too many individual pieces of work. Rather, it is better to estimate the amount of time it will take for a whole team to complete a particular task. Likewise, a group's estimates tend to be more optimistic than the individual ones. If a large group participates in a planning poker meeting, it will be more difficult to reach a consensus on the total effort it will take.

Stand-Up Meetings and Retrospectives

Stand-up meetings and retrospectives can be a very effective way to improve the effectiveness of your team. One of the key advantages of stand-up meetings is that they foster communication and synchronization between different teams.

Retrospectives are another important part of the agile playbook. While these are a helpful method to enhance teamwork, they can also be challenging. For example, practitioners have found it difficult to give and receive peer feedback. Peer feedback is typically given during meetings, but can also be provided through automated testing of code.

As a result of the COVID-19 pandemic, retrospectives have become more complex. This is because many teams have been forced to increase their frequency of these meetings, but are also faced with the problem of ensuring that the content is relevant to the work they are performing.

With more people involved in these meetings, it can be difficult to get all the members to contribute, especially when people are talking over each other.

Burn-Down Charts and Velocity - How to Use Them to Boost Your Rate of Work

Burn-down charts and velocity are both good indicators of project status. They aren't the only ones though. Using a chart to gauge your team's progress can be a useful and fun activity. It's also an easy way to track the quality of your work. A well-designed chart can be the best indicator of your overall performance.

Velocity isn't a magical elixir. While it can be used to improve your estimation and forecasting, it isn't a magic pill. There are several ways to boost your rate of work, including using the right tools, scheduling time for sprint reviews, and getting rid of the red tape in your office.

The key is to have the right mindset. If you're still looking for a magic formula, you may need to consider a new job title.

The old adage is true. If you're working on an important piece of software, it's important to be aware of what's going on with your product. Even if you're not in charge of the entire development process, it's important to have a clear picture of your current state. Keeping track of your team's output and their progress can help you make informed decisions about your next steps.

The burn-down chart is an impressive way to visualize your product's progress. By measuring and comparing the time it takes for completed tasks to be written, you'll see how fast you're churning out features, and how long your product is taking to go from concept to market.

Managing Scope and Change in Agile Projects

Managing scope and change in agile projects can be a very important aspect of project management. Without proper management, changes can spiral out of control and derail the project off the roadmap.

Changes to the project can be caused by different reasons. New data, new technology, changes in budget or resources, or a change in the project's schedule. In an agile project, a change in the schedule may lead to an increase in budget or a decrease in resources. This can affect the project's cost baseline, as well as the time and effort needed to complete the project.

When a change is made, it is essential that it is carefully documented and approved. If a change isn't properly documented, it can cause communication problems. It can also delay the project or make it harder to meet the schedule.

Using a project management system like Jira can help you manage scope and change. It allows you to add context to backlog items, as well as estimate and groom them. Having a clear definition of the project's scope is a good way to prevent scope creep. This is done by collecting initial information from stakeholders. Once you have gathered this information, you can begin to write the requirements for the project.

Leading an Agile Team

Leading an Agile team is an important role that requires specific leadership skills and a solid understanding of the core principles of the agile framework. The right people, in the right roles, and in the right environment can lead to high-performing teams.

Teams in the new Agile Way of Working are based on a self-organized, empowered, and collaborative working environment. Using a common approach to estimating work, decision-making authorities are able to guide their team members towards effective action. They are aimed at providing a sense of purpose and motivation for all team members. This is achieved by delivering tangible value and building trust.

In addition to a shared vision and mission, relationships within the team are built on trust. By constantly communicating and collaborating, the team is able to meet its responsibilities. Iteration goals, common Iteration Plans, and PI Objectives all help facilitate these interactions.

During iteration, a Product Owner works with the team to define stories and deliver them. A Scrum Master helps to remove impediments to progress and fosters a culture of relentless improvement.

These teams are driven by a commitment to delivering value to the end user. As a result, they are always exploring new methods and techniques to improve their delivery.

The Role of the Agile Project Manager

To be successful as an Agile Project Manager, you must have excellent interpersonal, communication, and time management skills. Without leadership, an agile team can deviate from the main goal of the project. The project manager helps the team work together and teaches the Agile mindset.

Another important role of the project manager is to coach the team. This is done to teach team members to self-organize. As part of the coaching process, the project manager works with the product owner to determine the project's objectives and goals. During this process, the project manager works with the team to determine the appropriate actions to take at the right times.

Often, the product owner leads the Agile project team. But, the Agile project manager will often be responsible for coordinating and demoing sprints. They also monitor issues and risks. By doing this, the team can be notified before they escalate.

Building and Motivating a High-Performing Team

Building and motivating a high-performing team requires the right mix of people and the right work processes. A high performing team should also have a well-defined vision and clear goals.

High performance teams are composed of skilled members who understand each other's skills, abilities, and working methods. They are also collaborative, engaged, and committed to the overall business goal.

As such, they report regularly and share their successes and challenges. This allows the rest of the team to know how they are doing. In addition, they are accountable to each other, which keeps the team on track.
When building and motivating a high-performing team, leaders should always strive to foster a team culture. This includes creating a collaborative atmosphere where everyone knows what they are expected to do.

One way to encourage teamwork is to allow team members to have input on a written team charter. The charter should outline what the group is responsible for, how they should work together, and what the company's objectives are.

Facilitating Effective Communication and Collaboration

To achieve business goals, effective communication and collaboration are essential. The lack of a good communication system can create a lot of conflict and frustration, and it can also prevent teams from working together efficiently. Another communication strategy is to encourage employees to communicate their concerns and ideas to their managers. When you allow this, they are more likely to get solutions to problems they have encountered.

A workplace that values respect can help reduce stress and boost productivity. Employees with more opportunities for feedback are more likely to feel valued. They are more likely to have constructive relationships with coworkers and improve their overall performance.

Team collaboration is one of the most effective methods of achieving business goals. However, it is not always easy to do. Communication challenges can arise because of different cultures, as well as poor leadership.

In today's business climate, research shows that over 80% of tasks require collaboration. Effective communication and collaboration can make or break a team.

Overcoming Common Challenges in Agile Teams

Often, an organization's Agile transformation can be hindered by communication issues. The key is to establish a clear mechanism for two-way communication, as well as regular meetings to discuss the latest developments and challenges.

A common challenge that organizations encounter during an Agile transformation is a lack of buy-in and commitment from management. This can be caused by the lack of a clear understanding of the process, or a lack of trust in the new processes.

There are many ways to address this. For example, using a community of practice to encourage a learning environment, or training for new managers to pass along their experience. Also, working with all stakeholders, including the executive team, to ensure that the organization understands its priorities in an agile context can help the entire organization move forward.

An effective Agile implementation requires a strong culture change. Companies that have adopted Agile in the past may be resistant to change, owing to legacy, resource constraints, or other underlying challenges.
To overcome this, organizations should first identify and understand the reasons for these silos.

Secondly, there are many ways to improve communication between teams. Daily stand-ups can help to facilitate these interactions. In addition, use intuitive tools to make sure everyone is up to speed. These types of communication techniques can also be used to keep employees informed of the latest developments in the project.

Finally, a shared mindset is often a blessing and a curse. If there is a shared sense of purpose among the various teams in the company, they will be more likely to engage and collaborate.

However, when there are inconsistencies between the various teams, there is confusion, and this can lead to issues.

Best Practices for Agile Team Leadership

Agile leadership is the application of principles and practices to build high performing teams. It teaches teams to work together to produce innovative software.

Teamwork is the key to delivering value to customers. In addition to a well-planned project, it is important to implement the right organizational practices.

These include leadership styles, cultural values, and communication cycles.

It's also worth incorporating the Agile Manifesto.

The Agile Manifesto outlines the main principles of agility, as well as the core values.

Essentially, the Manifesto describes how to deliver a working solution by setting up a common vision and building on the ideas of others.

A Kanban board helps an Agile team visualize workflow. However, it's worth noting that it is not always necessary to use one. You can also plan and visualize workflows using a project management tool.

One of the most exciting things about Agile is that it empowers team members to take imaginative leaps.

Instead of being limited by a predefined set of procedures, teams can choose their own processes.

Likewise, teams can adapt to changes in the market and respond quickly to customer feedback. By combining Agile's collaborative approach with lean project management techniques, teams can be more effective.

Agile Project Management in Practice

If you're looking to make an informed decision on how to run your next project, it's wise to take into account the advantages and disadvantages of Agile and traditional project management. The benefits can outweigh the disadvantages. However, not every organization is a good fit for Agile.

The agile approach emphasizes an iterative approach that integrates planning and execution. This helps reduce the risk of rushed testing cycles and costly mistakes. For example, an agile project is made up of small cycles - Sprints. These are defined by empirical data and are typically short in duration.

An agile project also entails constant feedback and collaboration. It's a good idea to have frequent reassessments of the work being done, as well as the overall progress of the project.

Aside from the obvious benefits of collaboration, the agile method focuses on rapid communication between team members. It's also a good idea to use tools such as email and in-person conversations.

Case Studies of Successful Agile Projects

Case studies of successful Agile projects can help organizations educate their executives and teams about the benefits of agile transformation. By learning from the best practices, an organization can improve its client experience, employee morale, and productivity.

For the ING Group, one of the world's largest banking and financial services firms, agility helped the company deliver products and services to customers more quickly. Since it began implementing agile methodologies, ING has seen increased productivity, employee engagement, and customer satisfaction.

One of the first steps to transforming an organization was to get management on board. ING leaders recognized the need to shift from a bureaucratic structure to a collaborative one.

In addition to re-focusing on quality and teamwork, ING also wanted to improve its customer experience.

Thermo Fisher recently combined content and agile marketing efforts. They created an agile marketing team and implemented an inbound methodology to better manage its marketing campaigns.

Chad's team has overcome several obstacles during the Agile transformation.

The biggest challenge was finding an objective way to prioritize work. To do this, he and his team had to move from traditional project management to an iterative process.

Despite their challenges, the marketing team at Sunlife understood the benefits of using Agile.

By bringing together cross-functional teams, they were able to focus on delivering value to clients.

Common Pitfalls and How to Avoid Them

Among the most common pitfalls are those related to the planning phase of research. Although planning is essential, strategizing too long can cost you opportunity.

This includes establishing basic criteria and priorities, as well as a simple template for the opportunity. Also, you should establish a basic threshold for risk.

Another hazard that can arise is poor performance reviews. It is imperative that you perform your duties as efficiently as possible. A lack of change control can also be a problem. Other pitfalls include a lack of a quality user help desk.

Finally, there is the publish or perish culture. This can affect a prospective author's motivation to contribute to science.

Nevertheless, a commitment to making a contribution will pay off in the end. Following your instincts will not get you into trouble, but if you do make a mistake, it is crucial that you learn from it.

Agile Project Management in Different Industries and Contexts

Agile project management is an approach to organizing work which can be applied in a variety of industries and contexts. It is a methodology that emphasizes collaboration, communication and responsiveness to change.

Compared to traditional approaches, agile methods can provide companies with a much faster way to deliver high quality, useful products.

It also promotes a more people-centric approach and encourages team members to consider themselves as integral partners in achieving your business goals.

Unlike traditional methods, however, agile projects do not use rigid, predefined schedules.

Rather, they are completed in short time periods called sprints.

The most important thing to remember about sprints is that each sprint is only one step away from a fully functional product.

There are many other benefits of Agile methods, such as reducing the risk of reworking your product during its development.

For example, if you find that a feature you have built is not working, you can easily tweak the design to make it better.

In fact, many pharma industries have started using the method to speed up their research and development, and to reduce the risk of a feature going wrong.

Other uses of the agile method include creating valuable content. For example, if you need to create a brochure, you can easily do so with the help of an agile method.

Advanced Agile Techniques and Strategies

Agile techniques and strategies to help technology organizations evolve their culture and boost product quality.

For example, they are designed to enable faster go-to-market times, increase customer retention, and improve team performance.

The best part about these practices is that they are highly flexible. Instead of trying to adhere to a rigid plan, teams can make changes according to client feedback.

In turn, clients can expect better results.

An online Kanban board, for example, can help teams visualize their workflow and track progress.

By using a collaboration platform, teams can eliminate barriers to communication. It also enables them to collaborate more effectively.

Other methods include refactoring and Continuous Integration.

These practices help avoid software rot and speed up the release of MVPs. They also allow developers to commit code several times a day.

While these processes are important, teamwork is the most important of all. Working together empowers teams to produce innovative software and take imaginative leaps.

To achieve this, agile teams use many tools and software applications. Some are used only by the team, while others are shared with other team members.

Depending on the project, an agile team may work with eight or more people. Generally, these teams are close-knit and share accountability for outcomes.

This is because the most effective teams are able to reflect on their behaviors and adjust accordingly.

Using a combination of agile techniques and strategies can help a company become more data-driven and smarter with its architecture decisions.

Future Trends in Agile Project Management

When it comes to future trends in agile project management, organizations are embracing new technologies and methodologies to boost efficiency, speed up delivery, and improve customer satisfaction.

The emergence of smart technologies and AI and ML-powered tools are accelerating product development and improving decision-making.

Organizations are now relying on a hybrid approach that combines traditional and Agile methodologies.

This practice helps teams synchronize and share information. It also promotes faster market response and enhanced customer service.

The technology is also enabling people to work together across greater distances. With the rise of mobile and video-based apps, people can communicate and work more effectively.

These technologies are making project management easier than ever.

The use of a collaborative workspace has been shown to reduce the costs associated with remote team members.

By using applications such as Slack and WhatsApp, team members can easily communicate.

The use of smart technologies is helping project teams to innovate and test faster. Using cutting-edge devices such as quantum computing and the Internet of Things can accelerate the development process.

In order to achieve this, organizations are investing in DevOps practices.

DevOps is a management technique that involves collaboration between teams to streamline development and deployment. Companies are combining these practices with Agile methodologies to deliver quality software.

Another emerging trend in agile organizations is design thinking.

This principle involves integrating customer needs early in the development cycle.

Design thinking has helped to create products that are more user-friendly.

Agile Project Management and Stakeholder Engagement

Stakeholder engagement is a vital part of any Agile project. A sound stakeholder engagement process enables you to realize the many benefits of your project.

If you do not, the results of your project may not satisfy your stakeholders.

For most projects, involving your stakeholders will make all the difference.

But, it is often overlooked. In Agile, this task is even more important.

Stakeholders are anyone who has an interest or influence in your project.

They include internal and external partners, executives, and senior level managers. As such, you should consider their inputs and interests as part of the planning and executing of your project.

Not only should you send out regular newsletters and emails, you should also schedule meetings to get feedback and inputs.

Some stakeholders may even have specific vested interests in your project.

This may involve the development of a formal charter or program, or a change of control.

For example, a continuous deployment strategy can help you measure the benefits of your incremental benefit realization.

Similarly, a well-developed and implemented Stakeholder Engagement plan will allow you to understand the issues affecting your organization and how to best address them.

A well-defined Stakeholder Management Plan will ensure that you do not overlook this crucial aspect of a successful Agile project. It is a vital component of your overall plan to achieve a successful, sustainable, and dependable Agile environment.

Stakeholder Engagement

When you're in the middle of building a complex project, it's important to engage as many stakeholders as possible.

You need their support to make sure the project's success isn't compromised by lack of attention from key players. Stakeholders include your internal team, external partners, and regulatory bodies.

A good way to determine which stakeholders are worthy of your attention is to do a stakeholder analysis.

This helps you understand who is interested in your project and how they can contribute. It's also a useful tool for identifying groups that may not support your efforts.

Another useful tool is stakeholder mapping.

It allows you to categorize your stakeholders according to their interests, power, and influence. Using this method will help you to maximize your project's impact.

Stakeholders are vital for building high quality software or a complex project.

Getting them involved can help you gain support, acquire resources, and reduce conflict.

The best way to engage these groups is to establish open communication channels.

This should include frequent updates to show progress. Having regular discussions about what a completed product looks like can prevent surprises.

Providing them with the appropriate tools and incentives is also a must.

Identifying and Managing Stakeholder Expectations

Identifying and managing stakeholder expectations is a crucial part of the project management process.

Having a clear vision of what your stakeholders want will help you to make smarter decisions and prevent scope creep.

Creating a communication plan for each group will also help.

A status update should be scheduled on a regular basis.

While you're doing all this, make sure that you record any activity that pertains to the project.

This will help you to understand how your stakeholders are feeling about the project, as well as to measure their performance.

Using Microsoft 365 for your project management tool of choice is a good way to keep track of all your projects, while also keeping track of your expectations.

In addition, it can be used to create project documents and keep your stakeholders in the loop.

When you're looking for the best way to manage your stakeholders, it's best to approach it from the perspective of expectation management.

To ensure that you're doing the right things, you'll need to identify your stakeholders, their needs and interests, and their abilities.

Communicating Progress and Value to Stakeholders

It is important for organizations to keep their stakeholders updated with the progress of their projects. They can use a variety of methods to do this. However, not all methods are appropriate.

For example, the most common communication method for corporations is through meetings. This is a good way to engage stakeholders, though some may not attend. Meetings also allow stakeholders to ask questions and contribute ideas.

Email is another popular tool. It makes it easy to deliver information quickly, and it gives proof that it has been sent. As well, email can help you monitor engagement.

Another way to communicate is through a newsletter.

Newsletters can be used for both project stakeholders and non-project stakeholders. Content can include information about recent announcements, policy activity, and the impact of the organization.

The newsletter can also include contact information and next steps.

Presentations are also a popular way of communicating with stakeholders.

These can be online, in person, or via video. Stakeholders can view presentations from home or work, and they can be sent as a reference.

A weekly digest can be a good way to share the progress of your projects. This can include a summary of the week's news, budget updates, and next steps. You can also incorporate photos, links, and other graphics.

A portal can be a great way to eliminate unnecessary meetings and provide real-time transparency. A portal can also be used to track project status and communicate with stakeholders.

Dealing with Conflict and Resistance to Change

As an organization grows and changes, people are constantly faced with a number of different challenges.

A common one is dealing with conflict and resistance to change.

The best way to deal with these issues is to first understand how these factors can lead to a conflict.

One of the most basic forms of resistance to change is lack of trust. This is a result of individuals being insecure about changes to the business. Another type of resistance is misunderstanding. It can also occur in groups.

In these cases, a group of people may share a common view of the situation.

Other signs of resistance include failure to add input and withdrawing interest.

These can be seen in individuals who believe that their opinion doesn't matter or who fail to take action in a new direction.

A new software system should be planned by focusing on user adoption.

If the new software is not planned through the lens of user adoption, it can be difficult to get people to use it.

Building and Maintaining Trust With Stakeholders

Building and maintaining trust with stakeholders is a vital step in a project's success.

Your communication style should be straightforward and clear.

If your message is unclear or ambiguous, you run the risk of misunderstanding or worse, offending a stakeholder.

In order to build and maintain trust, you should also be able to demonstrate that you can be counted on to perform when the time comes.

You may not be able to build a relationship from scratch, but you can maintain an existing relationship by showing commitment, listening and following through.

The best way to do this is to engage your stakeholders early. This allows you to shape the narrative.

Getting involved with your stakeholders at an early stage can help you avoid a plethora of mistakes.

Similarly, you should be willing to admit to making a mistake. Acknowledging your blunders is a good way to establish a level of respect and show your stakeholder that you are serious about the relationship.

The most successful strategy to build and maintain trust with your stakeholders is a combination of effective communication, good planning and a little ol' fashioned hard work.

Risk Management

Agile Project Management and Risk Management is a management technique that helps organizations to manage the uncertainty associated with projects.

It allows companies to define, plan, and implement projects. Moreover, it facilitates organizational success.

Companies that aim to achieve success should set up a novel infrastructure, restructure conventional roles, and integrate project decision process with business processes.

But before this can be done, senior management must understand the relationship between portfolio management and strategic project execution.

A project is a fundamental part of any company's business.

Regardless of whether the project is for internal or external purposes, its success has an impact on the organization.

Organizing for agility means developing a team of individuals who work together and independently.

This team needs to be led by a knowledgeable and experienced leader. The responsibilities of the team members should also be clearly defined.

They must be able to work across departments and disciplines, as well as handle changes in the project.

While organizations have long used traditional project management methods, agile approaches have emerged. These methodologies require shorter development cycles and allow teams to make changes. During the execution phase of an agile project, teams must make appropriate decisions.

To avoid delays, it is important to keep team members motivated and focused. Keeping them informed about progress and expectations is essential. By identifying and prioritizing potential project pathways, the team can determine the best course of action.

The Risk Management role

The role of risk management in agile projects is to systematically identify and mitigate risks.

This allows for informed decision making and prosperous anticipation. As an Agile team, you have to take the necessary steps to ensure that you are meeting the requirements of your clients and that your project will be successful.

Once you have identified the risks that need to be addressed, you should set up a risk register.

During these meetings, you should discuss the various risks and strategies that you can use to mitigate them.

For instance, you may remove a story from the backlog if it is deemed to be risky.

In addition, you can also use a risk modified Kanban board. This type of board displays the number of identified risks, their probabilities, and the response of the team.

To help you manage the risks that you have identified, you can create a risk review that is held on a daily basis.

Risk reviews should last about 30 minutes and should include a discussion among all stakeholders.

Identifying and Prioritizing Risks in Agile Projects

Identifying and prioritizing risks is a critical step in project management.

Although not all risk is bad, it is important to find ways to manage them before they become a problem.

Creating a risk register also makes it possible to see which ones are threatening your project.

While a risk register is a document you will have to maintain for a while, there are tools available to help you manage your lists of risks.

Identifying the best way to go about tackling the risk isn't as simple as throwing a risk management meeting once a month. Instead, it is better to set up a series of meetings with your team to discuss, evaluate and manage the risks that have been identified.

Developing and Implementing Risk Mitigation Strategies

Risk mitigation strategies can be a great tool for project teams to keep a project on track.

When used correctly, a risk management program can help an organization keep costs in check, maintain a good reputation in the industry, and keep internal and external stakeholders happy.

Developing and implementing a risk mitigation strategy requires collaboration and a comprehensive understanding of the risks and how they will affect the project.

A good strategy considers the impact of external and internal risks.

A risk assessment involves evaluating the level of the risk, determining its consequences, and identifying the resources needed to reduce it.

The process of risk analysis involves evaluating data, researching alternatives, and submitting a plan to a decision maker.

Developing and implementing a strategy is an ongoing process. It requires monitoring the risks as they change.

Also, it is important to test the plan to make sure it's working. Using strong metrics can also help you track the risks as they develop.

Developing and implementing a successful risk mitigation strategy can be a challenge, but it's an important part of any business.

Failure to properly manage business risks can cost a company valuable time and money.

Taking risks can be the best approach, but some are better left unavoided.

Keeping the risks in the forefront of stakeholders' minds is essential for an effective decision-making process.

Having back-up plans can also make promoting risk-taking less daunting.

Incorporating risk mitigation strategies into your product development program will help you keep your projects on track.

By starting early, you can identify and mitigate risks before they become major problems.

If you're launching a new service, be sure to solicit feedback from end users to help you address potential challenges.

Risk Assessment and Management in Agile Projects

Project risk management is an important component of Agile development. A project manager needs to have a plan to document risks and risk responses. This should include how and when these documents will be accessed. Ensure that all stakeholders are aware of what to expect from the document.

Risks in Agile projects are managed through several phases, including planning, analysis, and monitoring. During planning, a team of developers and product owners creates a prioritised list of new features. These are then decomposed into user stories. Each user story describes the intended behaviour of the system.

The team identifies activities that may be affected by the risks, such as shipping delays, manufacturing defects, or computer viruses. The team then assesses the impact of each risk and documents the probability of each risk.

As the iteration progresses, the team will perform a risk review and discuss the risks with stakeholders. They should also reflect on the project's state and discuss how to best address the risks.

Best Practices for Agile Risk Management

Agile risk management is a combination of minimizing and monitoring risks throughout the project.

This approach helps to keep projects on track and avoid delays. It is based on an empirical and flexible approach to software development.

In order to apply it effectively, it is important to understand the main risks.

The agile development process breaks down development into small, iterative releases.

This reduces schedule and performance expectation risk, while increasing the chance of delivering high-quality elements in a shorter timeframe.

During iterations, the team monitors the status of their risk register, keeping an eye on any new or existing risks.

These are then grouped into logical categories and assigned a rating and a probability score. A vote of confidence is used to determine which risks are worthy of further action.

The project's backlog is constantly being re-prioritized. The top five risks are mapped back to the requirement.

During this iteration, the engineering team works to address these risks. As changes are made to the backlog, the team is able to respond to and manage them in a timely manner.

Quality Assurance

Many companies are grappling with how to ensure the quality of their projects in an agile environment.

While traditional methods have been proven effective in the past, they are no longer sufficient to deal with today's fast-paced, unpredictable, and complex project management environment.

Fortunately, there are several methods and approaches that can help you ensure the quality of your projects in an agile environment.

The first is a predictive project methodology.

These methodologies begin by defining and documenting all of the product requirements and needs at the outset of the project. In addition, this approach focuses on monitoring the progress of the project to identify variance.

This allows the project to be adjusted to meet the product requirements.

An Agile project is defined by frequent interaction between the team and the client.

This helps the team to understand customer needs as they change over time.

Because of this, the team is encouraged to develop simple processes and solutions to meet the changing needs of the client.

Quality Assurance

Quality assurance has an important role to play in Agile project management.

QA engineers and testers should be involved in every aspect of the project, from design to implementation and maintenance.

A quality assurance team's role is to ensure the smooth functioning of new features, without slowing down the overall development process. Testers should also be encouraged to use tools such as automated testing technologies to accelerate their efforts.

The role of QA in Agile projects can seem daunting at first. However, with a little planning and time, the QA team can make a difference to the quality of your Scrum project.

A quality assurance team is a great way to help your Scrum team stay on track and on target for a successful sprint.

They can assist your team in clarifying goals, providing feedback about requirements and highlighting defects.

Besides, quality assurance is a key component of any software development process.

Having a well-informed team can help to minimize unforeseen rework and avoid costly mistakes.

To help the QA team to perform at a high level, you must provide them with the right resources, including a robust training program.

The Agile methodology requires a high degree of collaboration amongst all members of the team.

For example, the QA must play a key role in all review and retrospective meetings.

Having clear goals will also make it easier to identify deviations.

While it isn't impossible to perform any task in a QA-free environment, the QA team will still benefit from a collaborative approach.

Continuous Integration and Testing in Agile Projects

Continuous integration (CI) and testing in Agile projects are key steps in the delivery process. They enable teams to make small, frequent, and repeatable changes to the codebase.

This increases the velocity of the team, and increases the quality of the software as it develops.

It also increases the efficiency of the entire team, and eliminates a lot of the manual, mundane tasks involved with testing and integrating. By implementing CI and testing in Agile, a team can focus on more enjoyable, productive work.

Having a system in place to ensure that changes are merged back into the main code base quickly allows for greater visibility and improved communication. In addition, this method promotes collaboration.

Teams should also consider adding end-to-end tests to their continuous integration pipelines.

These tests will help them detect bugs and errors in the code.

During a test run, the automated test will identify issues, and can help avoid regressions in the working code. Adding a full suite of automated tests to your CI pipeline can speed up your build cycles.

Using a tool like Git to perform continuous integration makes it easy for teams to branch, merge, and track their code.

Additionally, newer technologies make merging easier.

Creating a system of automated tests can provide a fast, reliable feedback mechanism that can be used to detect problems. Ultimately, it will also decrease the cost of a project.

While CI and testing in Agile projects may not be the only way to get a working product out to your customers, it is one of the most important ways to achieve quality, speed, and agility.

The faster you can deliver your products, the more competitive you will be.

Best Practices for Agile Quality Assurance

Having a good QA strategy is essential to the success of any team. You need to plan ahead and use best practices to maximize your testing.

The best practice is to implement an automated QA solution. This allows you to get immediate feedback and prevents repetitive testing. It also saves on time and money.

Choosing a good quality assurance method is a critical factor for the success of any software project. By implementing a few of the following strategies, you can ensure that your product is as error-free as possible.

One of the more effective ways to improve the quality of your code is to perform regular code rewrites. This will reduce technical debt and help you avoid bugs.

Testers should be involved in all stages of the delivery process. They should be able to provide input on design discussions and stand-ups. Also, they should be able to participate in retros.

The QA team should identify and document scenarios that can be automated. They will also write test cases. However, the best practice is to make sure that they are written under the supervision of the development team.

A good testing plan should include the creation of a smoke testing script. These tests ensure that the code is not broken and that it is safe to test.

In addition, testers should use media files to illustrate bugs. Not only does this help them to catch them, it can also give developers more information on the environment in which the bug occurs.

Automated Testing and Test-Driven Development

Using test-driven development and automated testing in tandem can make coding easier. The two together can reduce your feedback loops to minutes rather than hours. Moreover, it can prevent regressions as your application grows in complexity.

Test-driven development is a popular practice for delivering a robust and maintainable software product. It is also a great way to improve your chances of success. You can also integrate it into your Continuous Integration (CI) pipeline.

Automated testing can save you time and money. Rather than running dozens of manual tests, you can use an automated tool to evaluate your functions without user input. Having your code pass an automated test can also guarantee that your product will be up and running after changes.

Automated test collateral is also useful in that it provides a more detailed view of the state of your code. It will also reduce the number of bugs inserted into your application. This will reduce your risks and lead to higher customer satisfaction.

Balancing Speed and Quality in Agile Projects

With the rapid changes happening in the tech industry, it is important to strike a balance between speed and quality.

It is also necessary to keep up with the newest trends in technology to ensure that your company stays ahead of the curve.

The key to agility is having a disciplined approach.

This means taking an organized, professional approach to development.

Using these techniques can improve the performance of your software team and deliver more than acceptable quality.

One of the key benefits of agility is that it provides fast feedback. Continuous delivery enables this.

When defects are discovered, your system automatically reacts. Tests are run quickly and errors are spotted at an early stage.

When you create products on a standard foundation, you are less likely to encounter errors. Also, your products are easier to adjust. These are great things to have, but they can come at a price.

Another aspect of quality is the design and architecture.

These components determine the ability of your product to adapt to future needs. Having a design that is well-crafted will ensure that you can quickly and efficiently implement new features.

In addition to architecture and design, the code of your software is also important.

The tests that you define can help ensure that your code is accurate.

There are many different types of tests, including unit, integration, and functional testing. Some of these tests can be automated, while others require manual work.

Resource Management

Agile project management is a set of principles and values that encourages teams to work faster, collaborate, and make changes. Projects require a plan to deal with changes and to deliver a high-quality product. In addition, projects consume resources, and they must be scheduled according to budget and time constraints. Having a good understanding of the components of a project management process can help you select the best one for your project.

A project management process can be anything from a traditional methodology to a more innovative method. Regardless of which method you choose, it requires a comprehensive framework to follow. Choosing the right framework can ensure your projects are successful and repeatable.

Many organizations have turned to the use of Agile to speed up delivery and reduce production costs. Agile project management methods incorporate a continuous feedback cycle and a more iterative process to achieve success. As a result, teams are more collaborative and autonomous.

Many common tools can be used to facilitate an agile project, such as a Kanban board. Some organizations utilize whiteboards, sticky notes, or even an online Kanban tool. These types of tools help teams to visualize and manage complex projects.

One of the most popular Agile methodologies is Scrum. It uses smaller teams that work on tasks in short time periods. Teams also use regular review and testing. This allows for responsive changes to be made at every stage.

The Role of Resource Management in Agile Projects

Resource management is a vital aspect of Agile project management. Without it, organizations can fail to deliver their projects. As companies shift from a portfolio management approach to a product-driven model, they must change the way they allocate their resources. In the transition, it is essential to invest in a system that enables organizations to efficiently manage their resources.

Resource planning involves identifying, analyzing, and allocating resources.

It is also necessary to keep track of the utilization of each resource and its productivity. This data is crucial to the organization's decision-making process.

The most effective way to do this is to establish a resource management plan within the agile project methodology.

By implementing this plan, an organization can ensure that it is allocating resources to projects that will bring the highest value to the company.

Developing a resource management plan requires an understanding of the individual roles of each team member. The team leader should collaborate with other members of the team.

They should work together to set up a resource management plan that incorporates time off and other planning considerations.

In addition, it is essential to maintain a skills database. This can be accomplished by hiring plans and training plans. Keeping people trained in their skills will help avoid shortfalls. Managing these issues can be difficult, however.

In addition, an organization must provide strategic and financial independence to its Agile teams.

Using digital dashboards can help with this process. These tools can support the organization's strategic plan and address organizational capacity, risk exposure, and talent.

Prioritizing and Allocating Resources in Agile Projects

In agile projects, the resource allocation process is crucial to ensure that the right people work on the right tasks. The allocation process takes into account the skills and experience necessary for each task. Moreover, it can promote efficient workflow between team members.

The first step in prioritizing and allocating resources in agile projects is to define the project objectives. These include defining the project's importance and setting priorities for each task.

During this phase, a Gantt chart can be used to identify the tasks and their relationship to other tasks. For example, a task that involves programming may be scheduled for the same time as another task that is based on data collection. Identifying and scheduling the correct tasks will save time and avoid potential conflicts.

Allocating resources in agile projects requires a different approach than traditional planning methods. In agile methodologies, there are frequent requirement changes, and understanding the supply of resources is vital.

Using a Gantt chart or other project management software can help a project manager identify and schedule the right tasks. Once a task is identified, the project manager can communicate the schedule change to all team members.

Using software can help keep track of how many hours are spent on individual tasks.

This information can help the project manager prevent overallocation.

Managing and Optimizing Resource Utilization in Agile Projects

Resource management is about making the most of your team's time, skills, and resources to complete your project.

Resource management can help you increase your project's performance and prevent burnout. Resource allocation reports can provide a high-level view of the availability of your resources, and give you the information you need to avoid delays. The key is to keep track of your workloads to ensure your team is working at their maximum capacity.

A good resource plan ensures a fair rate of resource consumption. Having an accurate picture of the work your team is doing helps you to prioritize and complete the most important tasks first. If you find you're not utilizing the resources you have, you can re-allocate them.

Resources are the backbone of a project, and resource utilization is an essential part of managing your team. Resource managers are responsible for allocating resources and tracking their status.

One of the best ways to monitor and optimize resource utilization in an agile project is to use a resource management system. There are hundreds of programs available that will allow you to do this easily. It's also a great way to create a central pool of resources.

This will let you track your inventory, financial, and material resources.

Best Practices for Agile Resource Management

The best practices of agile resource management are important to successful projects. They include the ability to distribute expertise among teammates, efficient communication, and good chemistry.

The best practices for agile resource management also help to increase team productivity.

In order to plan resources, you need to look at several factors, including the amount of work each person can do in a week, the amount of time each person can take off, and the time it takes to finish each project.

You will also need to factor in the availability of team members.

Resource management can be a little complicated if you have multiple Sprints in your project pipeline. But with the right tools, you can find out accurate information about your capacity and utilization.

You can also use a digital dashboard to track your organization's resources and address risks.

For example, you can create dashboards that address external workforce factors such as the number of hours per day or week that workers are expected to be in the office.

Advanced Resource Management Techniques and Tools

The goal of resource management is to maximize utilization and avoid burnout. When resources are overused, it can cause delays and even lead to stress.

These issues can result in budget overruns and other complications.

In order to minimize cost and minimize stress, your resource management solution should be able to provide you with a global view of your team's capacity.

This is particularly important for businesses of all sizes.

You can also implement resource leveling to make sure that no resources are overused.

You can also do this when you're sharing resources with other projects.

This allows you to adjust the start and end dates of your projects.

Best Project Management Tools

Now that you know the different benefits of project management tools, let's look at the best Scrum and Agile tools available on the market these days.

Jira

One of the most popular Agile tools available on the market these days is Jira. It was created to help project managers and their teams track, plan, and develop products that fulfill the client's expectations. The users of this tool have an option to create user stores, design, and also distribute the tasks. Doing all this helps prioritize teamwork, making it perfect for an Agile team. Since all the discussions that take place are visible to anyone using it, transparency in communication also improves.

Jira is packed full of excellent features, such as scrum boards that can be fully customized, custom filters to create backlogs, and visual representation of the reports and documents produced. It will come in handy if you are looking for seamless implementation of Agile methodology. Jira also has multiple functionalities, and some of its key features include efficient management of backlogs, customizable Scrum boards, Sprint management, time and task tracking, and presentation of progress reports.

Ntask

You can use this software to remedy any problems associated with project management. This is also an incredibly versatile tool. Even though it's a project management software, it offers different features that make the implementation of Scrum and Agile quite easy. This tool does not have any add-ons; therefore, all the features it offers can be utilized from the beginning. If you are looking for a tool for the efficient transfer of information among team members, then its meeting management features will come in handy. Once you start scheduling regular meetings using this software, the team's overall productivity will automatically improve. It offers helpful features such as task and risk management, team collaboration, portfolio management, multiple views of the scrum board, a risk matrix, and the team's activity management.

Monday.com

This is a very good choice if you want to start managing different projects within a team.

Different features are provided by it, including customizable notifications and automation that ensures the team is focusing on all the important tasks instead of getting distracted.

The tracking capabilities offered by it can be used to track the progress the team is making along with the progress made on different tasks associated with the given project.

The visual representation of tasks and the roadmap for the team can also be created and shared using this tool.

Creating product backlogs becomes quite easy once you start using them.

It also offers different sorting options. Some key features that it offers include easy product backlog, Sprint planning, the seamless process of workload management, task dependencies, prioritizing project management, effective collaboration, and is an intuitive and easy-to-use interface.

Clarizen

You can start taking care of the enterprise's requirements, not just the team, using this wonderful platform. It's essentially an automation software that will simplify professional service processes and project management.

It offers incredible scope for work integration needed for improving a team's overall productivity and efficiency. The time-saving workflows it offers can be used to ensure that the projects are developed quickly and efficiently.

If your team usually undertakes any projects based on recurring strategies, then this automation tool will come in handy.

If the team usually depends on third-party applications, they can all seamlessly integrate with this platform.

Some other helpful software with which you can integrate this tool is Jira, Salesforce, and SharePoint. Multi-tiered network security, risk management, and assessment, efficient time tracking progress and scope management, creating and tracking budget, security and encryption, resource management, and social collaboration are some of the features offered by this brilliant agile project management tool.

Yodiz

If you are looking for a comprehensive tool to easily implement Scrum, then you should consider this option. It has various features that help with the seamless implementation of Scrum protocols. You can use it to improve the overall productivity of the team. User stories, releases and backlogs, and sprints can be efficiently managed using its toolkit regardless of the project you are working on. Different feature requests can be tracked by using its management module. The team will better understand what it needs to focus on once you start using the priorities feature for listing the tasks. Some of the key features it offers include burndown charts, issue tracker, third-party application integration, product backlog management, and Sprint management.

Scrumwise

Another tool that's quite helpful in implementing the Scrum framework is Scrumwise.

It's an extremely simple-to-use software that offers specific features based on the Scrum framework to improve your team's efficiency and productivity.

Its simple interface, whether in terms of creating backlogs or setting custom filters on the product backlog, makes it easy to use.

You can also break down any work-related activity into smaller and more manageable tasks, along with the creation of a focused checklist.

Since individual team members will be performing individual tasks, you can use the tracker application to track their performance and manage the estimates.

This is a multifaceted tool, from third-party application integration and real-time tracking to product backlog management, team chat functionality, and task management.

Agile Manager

This tool is created by HP and essentially helps organize and guide the Agile teams from the beginning of the product development cycle until they have created a fully deployable working code.

During the initial stages of the cycle, especially the release plan stage, managers can use it to gather user stories and decide how the team will tackle them.

This step by itself sets the stage for subsequent Sprints along with deployment.

During the Sprint, Scrum masters, along with developers, can record the progress they've made on user stories allocated to them and bugs in them.

The dashboard provides charts plotted with progress and any failures the team encounters. This, in turn, offers the needed insight to take corrective action.

Active Collab

From generating bills and dealing with multiple tasks to tracking the time taken, this wonderfully organized software helps development teams deliver code and accounts for the time taken. This tool helps create a list of tasks that are assigned. It also offers the feature of tracking how the tasks are completed from conception until delivery. It also has a system-wide calendar that the teams can use to understand and track everyone's roles. Since the system automatically checks the time devoted to each of the tasks, the team members can then determine whether their estimates are accurate or not. It also supports a collaborative writing feature, which means the entire team can work on documentation together.

Agile Bench

This is essentially a hosted platform ideal for tracking the work assigned to individual team members.

The release schedule usually starts as a backlog of all the user stories along with any other enhancements.

As and when the tasks are assigned, the team needs to determine the cost of development by estimating each task's complexity and its business impact.

The dashboard of this tool helps track both these metrics so that the team members can easily determine the important tasks. It also shows when a specific member is overloaded or overburdened.

Pivotal Tracker

This is one of the tools offered by Pivotal Labs that support Agile development. The core of any project development consists of a page listing all the tasks expressed as user stories.

The team members can rank the complexity of the tasks with the point system provided by Pivotal Tracker. This tool also enables users to track the tasks that are completed every day.

Telerik Teampulse

Telerik is quite popular for its different frameworks for app creation in the mobile marketplace. All the experience they have obtained by doing this has been bundled into creating their own code, known as teampulse.

This Agile project management tool helps track projects.

The primary screen shows a list of all the tasks to be completed. It helps track the progress made by the team as they start completing the tasks.

Different configuration options are also provided in the menu, including various reports to determine how the project is progressing towards completion.

If your organization uses any of the other tools from Telerik, Teampulse can be easily integrated.

Versionone

If an enterprise wants to opt for Agile project management and development, it will require customized tools to juggle multiple teams involved in multiple roles. This needs to be done because all the teams will ultimately be working toward a common goal. VersionOne is designed to organize different groups involved in a department across the organization by offering a stable platform for communication where everyone can actively take part in planning initiatives and for creating continuous documentation. It offers Kanban-like boards that can be used to track ideas and user stories throughout the development process until it is transformed into a properly working code. This system tracks all Sprints and organizes the Sprint's retrospective analysis before a cycle starts. This tool also has an in-built code that ensures it can be easily integrated with other packages.

Planbox

Planbox is a wonderful Project management tool that offers a separate level of organizational power when multiple people need to work together to achieve a specific goal.

The top level consists of initiatives that are usually the biggest or most important tasks to be achieved.

This contains projects that include built-in items which are filled with all the tasks to be completed.

When the team finishes a specific task, this tool tracks the progress on different levels and offers reports to the stakeholders. An interesting feature of this tool is that it enables you to keep the customers in the loop so that they can provide feedback before the code is finalized or set in stone.

It also has a time tracking feature that allows the team members to compare the time spent on an item against its estimate.

It can also be integrated with other tools for storing code tracking, customer satisfaction, and bug tracking.

Leankit

This project management tool imitates the whiteboards in a conference room. This is where all projects start.

You and your team can use this tool to post virtual cards or notes representing all the tasks, bugs, or user stories to be addressed.

The board is updated when the team finishes a specific task. Well, it will update itself quicker than any other manually maintained whiteboard.

You can also use this software across multiple teams that are working together in separate spaces. This further improves collaboration within the team and across teams.

Axosoft

This project management tool helps track the project in three ways. Using Release Planner, you can obtain a tabular bird's eye view of the tasks, bugs, and user stories.

The different entries can be assigned and marked as finished by using the simple drag-and-drop option it provides. You can use the burndown charts to check how the team is performing in relation to its goal.

This information is provided graphically, so you don't have to skim through paragraphs of data.

To ensure that all the team members are on the right track, the prominently displayed projected ship date helps. This tool also offers a Kanban-style card view where different cards can be used to represent different tasks.

Another wonderful feature this tool offers is its customer portal.

You can use this to obtain customer feedback about the development process, designs, etc.

Agile Project Management and Governance

Agile project management and governance is the practice of managing projects with a flexible and customised approach. This enables more effective collaboration between project teams and end users. In addition to this, middle managers are also crucial to project governance. They provide oversight, ensuring the quality of software outputs and maintaining an agile environment. Unlike traditional project management, the use of agile methodologies in ASD projects is often governed by a set of principles. These include working with a set of values and decomposing solution requirements into increments.

Traditionally, ASD projects follow a set of stages of feasibility, design, and building. During each of these phases, the customer defines the solution's objectives. After each iteration, the customer will send feedback on the project's progress.

The TECHCOY agile project team had several sub-teams. Each team was cross-functional and co-located. Several meetings were held a day to coordinate with each other. One of these was a monthly project review meeting.

The agile project team was led by a divisional CEO and consisted of a team of junior-level developers, a Scrum leader, and three MMs. Throughout the project, the MMs acted as gatekeepers, coaches, and pastoral care providers. They were instrumental in enabling the team to deliver the best possible outcomes.

Governance in Agile Projects

If your organization is considering implementing an agile project, it's important to know how to govern it. Good governance is about more than a Gantt chart. It also entails the judicious use of empirical performance metrics. These measures should help you keep the team on track and improve the overall quality of your work.

The first step in developing an effective governance model is to clearly define your objectives. You need to understand what value your project can provide to your business. For example, you should identify how much revenue it can generate. This information should help you understand if your project can be approved or if you should send it back for rework in a previous stage.

To make sure you get the most out of your agile project, you must set up the right processes to guide it. There are several different types of governance you can choose from. Each type should be designed to fit your organization's context.

One of the most important features of agile governance is its simplicity. It is not about over-engineering the processes. Instead, it is about providing the right authority to your team.

Agile teams should be empowered. This means giving them the tools they need to operate effectively and report on their progress. In addition, you should give them the power to make decisions.

An effective agile team should be able to deliver working features in two weeks or less. They should be able to take quick corrective actions, too.

Establishing Governance Frameworks and Policies for Agile Projects

When it comes to establishing governance frameworks and policies for agile projects, there are several important things to consider.

First, it is critical that teams have clear, measurable performance and governance metrics.

The second is to establish common objectives and goals.

Next, define the key points of contact for issues. Using these guidelines, project managers can identify who should be involved, and how to resolve issues.

Finally, it is important to implement an appropriate level of oversight. Several methods are available to achieve this.

Some include internal audits, rule books, and processes.

As your organization moves towards Agile, make sure your governance is based on best practices.

These include a clear definition of expected outputs from the projects, a well-defined and defined process for decision making, and an efficient way to share information.

Another important factor to keep in mind is the ability to change your governance model. Changing your approach requires a lot of patience and consensus building.

It is also crucial to ensure that your new approach is not only practical, but reflects the reality of the current environment.

Best Practices for Agile Governance

The best practices for agile governance are those that empower and energize delivery teams to deliver value.

Rather than command and control, an Agile approach empowers teams to make decisions and implement them.

An enterprise-level product backlog can be useful for this. It provides a high level view of the major components of a project, including its objectives and goals.

This allows the leadership team to identify any critical gaps and ensure the right people are working on the right projects.

A collaboration platform can streamline the process of organizing work.

In addition to removing barriers to communication, this type of platform can help build trust and confidence among the organization's leaders.

In fact, a well-developed process is an important component to any Agile implementation.

Although the process is not limited to software development, a hybrid approach is the best way to acclimate organizational leaders to the Agile mindset.

When selecting a tool to support an Agile project, consider a platform that supports visual documentation.

Providing simple, readable project information, such as a burn down chart, can help senior management and other stakeholders track progress.

An effective Agile process should also be able to manage change. To do this, leaders should analyze the value of their Agile projects and consider how to properly manage the risks.

As the project develops, they should provide additional control processes to help maintain and enhance quality.

Another important element is a collaborative workspace.

While team members will inevitably disagree on certain aspects of the project, they should learn to work together for the good of the project.

Advanced Governance Techniques and Tools

The demand for advanced governance techniques and tools is growing. Regulatory involvement, consumer demand, and the rise of newer business technologies are all reasons to address this challenge.

A leading global retailer invested in an enterprise-wide analytics transformation. In doing so, they established data domains and a data management organization (DMO).

They prioritized deploying transactional data and product data.

Organizations should also incorporate a data quality program into their governance efforts.

Depending on the type of data, a company may need to implement security tools.

Immuta offers a suite of features that can help ensure data integrity.

Another option is to create a centralized metadata repository.

Apache Atlas supports integration with various Hadoop components and provides an audit trail for changes in metadata.

The Future of Agile Governance

The future of agile governance will be a combination of digital, social, and human innovation.

Governments will need to become more responsive to the needs of the people they serve. In turn, this will help them meet public expectations and improve resilience.

AGC aims to provide a roadmap for effective agile government in practice.

It brings together governments, nonprofits, and academic institutions.

AGC has released a significant body of work since its creation. Some of the most notable include The Road to Agile Government, Adopting Agile in State and Local Governments, and The IBM Center for Digital Government's Human Centricity in Digital Delivery.

For the most part, however, the best approaches are disciplined and integrated.

They involve an organizational mindset that supports change and experimentation.

To implement a true business agility strategy, senior leaders need to develop a new set of capabilities.

These capabilities include a vision, a strategic plan, and a way of doing things.

For example, an organization should identify how they currently operate, and the technology they use.

They should also make sure that their current processes are flexible. This will allow for faster feedback loops and more frequent changes.

The best approaches involve integrating the most logical and efficient solutions into the core procedures of the organisation.

As such, it's important to document what's working and what's not.

Taking advantage of technologies such as smart contracts and collaborative networks will also be key.

Not only will they enhance and optimize the effectiveness of the operations they support, but they also bring immense value to the organization.

Common Issues with Project Management

The Project Management Institute Research (PMIR) estimates that $122 million is wasted for every $1 billion in projects in the United States due to a lack of project performance (Project Management Institute [PMI], 2020).

When managing a project, you are trying to minimize and contain the risk associated with it.

A project might take too long or be stopped abruptly.

In other cases, it may not deliver the promised results or come in at budget. There are several factors responsible for these failures.

When planning your project, you must account for every failure factor that can decide the fate of the project, which include:

- Poor communication between team members.
- Disunity amongst team members.
- Unrealistic expectations and projections.
- Poor planning and execution.
- Incompetence on the part of the project manager or team members.

You may have noticed that many of these are within your control. Some of them are not; every once in a while, two or more professionals can get into such a disagreement that they can no longer work together.

However, for the most part, these issues can be prevented through proper planning and communication, along with good leadership.

Remember the 7 P's of the British Army:

> *Proper Prior Planning Prevents Piss-Poor Performance*

Knowing the plan ahead of time is key to success. If your team members are adequately prepared for their tasks and everyone is communicating well with each other, you will be unlikely to face a project failure.

Who Can Be a Great Project Manager?

Leading a team means that anyone can be a great project manager as long as they have leadership capabilities.

Think about your own strengths

Are you good at communicating with all kinds of people?

Are you strong in your ability to manage conflict between others?

Can you envision things from a bird's eye view?

Are you a good coach?

Can you motivate people?

Is it easy for you to see which items to prioritize?

Are you adaptable to change?

It's rare for one person to have all these skills, but anyone or more of them will help you lead successful projects as a project manager.

Now that you've read this entire book, you should have gained a solid grasp of what agile and project management are all about.

When you think back on what you've read, what character traits do you have that work well with what you've learned?

What values do you have that fit right into the agile methodology?

Considering these questions will help you become an ideal project manager.

In the next book, we'll talk more about being an effective one, but for now, see how your personality fits into the lead role, no matter your education or current title.

Improving Your Skills as a Project Manager

You will want to be the best project manager you can be, so here are some ways you can improve your leadership skills.

Many of these items are either enhanced by doing more and more projects, so you would learn and improve naturally over time, or by reading how others handled similar situations and succeeded or failed. A strong work ethic and desire for improvement will take you a long way.

Keep projects simple

What's the simplest way to do a task that will achieve the result you want? Learn to look for inefficiencies.

You and the team should always be asking yourselves if there's a better and more efficient way to do a task.

Most people think that overcomplicating things will ensure the best results but the opposite is true!

Great achievements are usually the result of being focused on one simple idea and eliminating the waste around it.

Continually improve project planning

The more projects you manage, the better you'll become at planning them.

Write whatever lessons you've learned from one project down, so you can have them for the next, and building on them as you go.

Control projects: maintain budgets and timelines, limit scope creep

You'll be working on budgets and timelines with your team, so the experienced members will be able to validate whether the plan is too ambitious or not.

The great thing about agile project management is that the iterations are relatively short, so you'll find out quickly if the plan is too ambitious.

Once you do, you can then correct it immediately.

Scope creep is something you'll probably face from both the customer and your team while they discover new things during the project's process.

It's a matter of you being able to hold the line and always think about what was initially promised.

Improve risk management

You can read about risk management to get started as a project manager, but this is another skill that you'll improve over time, as long as you continue to learn through previous projects.

You'll have a better grasp of potential risks that can pop up, so your mitigation plans will become more effective over time.

Learn to be a good diplomat

Managing conflict falls mostly on you as the project manager.

You'll probably have issues within the team and between the team and the customer.

It's natural to have disagreements. You'll also need to build trust with the customer.

Allowing both sides to be heard and finding a solution is key.

Customers might get upset that the team isn't working exactly the way they think it should, or they may be pressuring the team in a way that makes the team less efficient. As a project manager, you'll need to take care of these situations without ruffling too many feathers on either side.

You'll learn this on the job, but you can practice outside it as well!

Do you belong to a sports league, or maybe the PTA at your child's school? Volunteer somewhere?

If you belong to any kind of group, take the opportunity to work on being diplomatic whenever you can. If you look for opportunities, you will find them!

Communicate changes to the team as soon as you can

The team needs to know ASAP when things change, so they can prepare themselves.

It may be a change in the stakeholders, the technology, whatever.

Set appropriate expectations and abide by them

Make sure the team knows what you expect of them; they should understand it correctly, so there's no room for miscommunication and misunderstanding.

If a team member isn't doing what they're supposed to do, you will need to call them out on it, though not necessarily in front of the whole team.

Organize your space and work

Whether you find tidying up fun or not, it is a fact that an organized workspace makes it much easier for the work to get done.

In a tidy workspace, no one has to spend time looking for the resources they need, and it's much more calming to come in to work at a clean desk rather than a disastrous one.

Similarly, organized work makes life much easier.

Everyone knows what to do and when to do it, and there are no questions about what comes next or where to find any needed resources.

Be creative

You know APM is all about flexibility and adaptability, so being creative would be a vital trait of a good project manager.

You can improve this outside the office by doing whatever creative projects bring you joy: crafting, woodworking, etc.

You can also get creative by changing up a habit, like the route you normally drive to work in the morning.

Drive in a different way and notice what's changed. You could also spend more time outdoors or listen to or create your own music.

Think critically

Being able to make decisions in the face of uncertainty and tackling problems confidently are also important for project managers.

If you're not already in the habit of thinking this way, there are some games and articles that will help you.

The three habits that will help you improve are questioning assumptions, reasoning through things logically, and looking at the situation from a different viewpoint.

You can practice these anywhere.

When you're reading an article online, find the assumptions, then question them. Is the argument logical? What would be the opposing viewpoint? The more you practice—as with most skills—the better you'll become.

Conclusion

Every project has its uncertainty and no count which methodology you select there will be some degree of unpredictability, specifically early on in a project.

Agile methodology which depends heavily on group coordination, conversation can assist you set expectations and manage that uncertainty.

Over the years groups have carried out agile techniques to varying degrees, to expand pace to market their product, extend productivity, boom strategy, improve operation effectiveness processes, improve product satisfactory etc.

Many groups that are new to agile process seem to conflict with estimation.

It's important to clear the air about Agile estimation and how it can supply results if applied in a proper way.

Estimates can assist an company to set a purpose and expectations about what a team can deliver, however due to lack of terrible planning groups struggle with estimation.

In many cases, it has been observed that things turn out to be absolutely rough if matters are estimated barring appropriate planning and perception that may in the end jeopardize the total project.

It's a confirmed fact that to make a precise prediction in a venture you want to have a correct estimation.

There are lot of matters that need to be taken into consideration while calculating budget in any project like infrastructure, time frame, man hours etc. to get an concept of a design and a purpose to measure against.

Once in a challenge you set a goal, teams want to come up with plans and estimate tasks.

Approach that has been widely used in agile methodology to correct flaws in estimation strategies is planning poker.

Planning poker concept every now and then called Scrum poker is a simple however powerful system that corrects any false precision and makes team-estimating faster, more accurate, and more fun.

However it has a 'con' side to it.

The estimating memories with Scrum poker idea is based totally on the story's complexity.

A story for example with number 5 can be extra cumbersome to whole than one it really is has a range 3, but it doesn't supposed that the 5 will take extra time than 3 to complete.

Estimates that are completely based totally on time can now and again make planning commitments complex and uncertain.

Made in the USA
Coppell, TX
12 May 2023

16752731R10057